D0279819

GOAT KEEPING

VALERIE HITCH

INTERPET PUBLISHING

Promoting responsible pet ownership

Published by **Interpet Publishing**
Vincent Lane,
Dorking,
Surrey
RH4 3YX
England

ISBN 978-1-8428-6221-6

Editor: **Candida Buckley**
Designer: **Sue Rose**
Photography: **Sidney Jackson**

Production Management:
Colin Gower Enterprises Ltd
Printed and bound in China

THE AUTHOR

Born in the year of the Goat/Sheep many Chinese moons ago, I was destined to keep goats. Animals have been a big part of my life and I have had a deep empathy for all creatures since childhood. Three years after I married my husband, we moved into a very rural area, and in 1983 we began keeping sheep. Five years later we took on two unwanted goats. In time, we had a small private santuary. I now advise concerned goat owners about their animals, using my knowledge and years of experience, which I have distilled into this book.

Contents

Introduction

In The Reader's Digest Encyclopaedic Dictionary of 1964 the definition of the word GOAT was as follows, and I quote; "Hardy lively wanton, strong-smelling usu. Horned and bearded ruminant quadruped (genus Capra); zodiacal sign Capricorn." Now forty plus years later many people still perceive goats as being dirty smelly and bad tempered. In fact, this is far from the truth, as any caring responsible goat keeper will tell you. The reality is that goats are intelligent, sociable, affectionate, people-loving creatures who do not smell (unless they are entire billies). If they are cared for correctly, and allowed contact with their own kind, they are great fun and will give you endless hours of entertainment in return for your loving attention to their every need. They will indeed require a lot of your time, but association with them is very therapeutic and enjoyable.

Goats are intelligent and relate well to humans.

GOATS' LONG TIME ASSOCIATION WITH MAN

Goats are known to have been one of the first animals domesticated by neolithic man when he gave up hunting and became a farmer around 3,000 to 6,000 BC. Goat remains found by archaeologists have been dated as being between one and eleven million years old, which gives us an idea of how their intelligence, mental and physical agility has enabled them to survive by overcoming the rigours of climate and terrain by utilising whatever forage has been available, whether it be on a rock face, up a tree or in an overgrown forest.

Man has kept goats since Neolithic times.

Some archaeological remains indicate that the first domestic goats were kept in countries such as Iran and Iraq, travelling with their owners when they migrated into the African and European continents. Goats are now kept throughout the world in widely varying climates, providing milk, meat, hair, and leather for their owners, and are much valued. Indeed, recent schemes to provide a goat to families in Africa are enabling people to make a living by selling surplus milk and offspring, and allowing them to buy other commodities (such as grain or seed for vegetables) with their new income.

GOATS AND MYTHOLOGY

Several world mythologies feature goat figures, indicating the importance of goats to early civilizations.

In Greek mythology Pan, the god of flocks and herds, is represented as half man and half goat in shape, having the animal's legs, ears and horns. Pan was also revered as the god of fertility. In time, he also came to be regarded as the personification of Nature itself

In German mythology, the Teutons believed that two male goats pulled Thor's chariot, as he drove across the heavens in a thunderstorm. While in Christian mythology a goat represents the devil himself. Occultists have portrayed Goat males in particular as having supernatural powers, thus adding to the undeserved myth that goats are evil creatures.

In mythology Thor's chariot was drawn across the heavens by goats in a thunderstorm.

GOOD GROUNDWORK FOR GOAT KEEPING

The advice in this book will show that if your preparation and homework are good you will obtain a great deal of enduring pleasure from keeping goats. They may live to an age of ten or fifteen years; and so they will be with you for a long time if you purchase very young ones. (We have had them until seventeen and even twenty years of age.) Rescue goats are often available for new homes, so it may also be possible to take on older animals and enjoy their last few years. It can be a very rewarding feeling as you gain experience from "the old hands." The most successful way of keeping any animal is to do your homework first; find out about the animal, in this instance, the goat, its history its requirements, food, housing, daily care etc. Getting it right from the beginning with give your relationship with your goats the best possible start, and ensure that you enjoy each other's company right from the start. Even if you were born with a natural empathy for animals, you should always be prepared to learn more about your goats.

Goats can survive on a wide range of herbage material, which would not be so well tolerated by cattle or sheep. But contrary to belief, they will not eat everything. They will almost always nibble at things to "sound them out" but will not always consume them. It is, however, important to keep polythene and other indigestible material out of their reach (as a precaution) and to learn which plants are poisonous to goats.

Planning Your Herd

ANY LAND ON WHICH GOATS are kept is classed as agricultural and you will require a "Holding Number" for that premises. When animals are brought onto that property, or are moved onto another, the Holding Number will have to be entered somewhere on the Movement Licence, this is a legal requirement. Records of movement and animal body disposal must both be kept up to date. Over time you will receive a great deal of literature from DEFRA (the Department for Environment, Food and Rural Affairs) some of which will not apply to your situation. Any goat brought onto your property should be carrying at least one if not two tags in its ear identifying it individually, and its place of birth.

In England and Wales, you will also need to be familiar with the Animal Wel-

A garden shed style building may be a budget conscious option for housing.

When planning your goat herd make sure you have unencumbered access to grazing land.

fare Act 2006, which states the required needs of any animal to be as follows:

A need for a suitable environment
A need for a suitable diet
A need to be able to exhibit normal behaviour patterns
Any need it has to be housed with, or apart from, other animals
A need to be protected from pain, suffering, injury, and disease
Any other need of that animal

Is the land yours and near your home? Rented property could place you in a difficult position if the owner decides to claim it back. Keeping goats out of sight from your home, or somewhere that involves travel can present problems. In bad weather, you need to be sure that you will be able to get to the animals on a daily basis.

Is the land securely fenced? If not, how much will it cost to do this? Goats are notorious escapologists. Does the land have any poisonous plants or surrounding bushes that require removal?

Make enquires into the cost of suitable housing. Custom built goat houses are available, but may prove expensive. A good garden shed may be a more budget-conscious option. It will need reinforced inside walls. If a handyman is available, perhaps a goat house could be custom built to your specifications, guided by this book.

You also need to check that there is a local veterinary practice that will deal with large animals rather than just the family dog or cat. Also, do you have access to transport to take your goats there? Vet call out fees will be high. However, there may be emergencies that will require this service.

You also need to establish where you will be able to buy your feed, and supply your animals' straw and hay requirements.

You should also ask if there are people who would look after your animals if you go on holiday or have to be away for a day? Goats have to be checked at least once a day after being let out to ensure that all is well.

It would also be a good idea to sound out your near neighbours to find out how they would feel about your having goats, which can be noisy at times.

A LIST OF EQUIPMENT

You will need the following equipment:

Wheelbarrow
Pitch or garden fork
Yard broom
Dustpan and brush
Rack
Garden gloves and disposable gloves for medication events
Grooming equipment: brush, and rake-style comb
Foot clippers and rasp
Collars (best not left on permanently)

A selection of tools which will make goat keeping easier.

and leads. These are for tethering, and taking your goats for a walk.
Feed bowls
Field hayrack. This will encourage animals into the fresh air and is necessary for animals who don't have access to main goat house.
Interior hayracks. These can be in the individual stalls or a single hayrack, large enough to feed all animals may be used, if a communal management system is used. All your goats need to have access to the hay, and you will need to be aware that there will always be a greedy bully.
Buckets for stalls and/or goat house
Flea and mite treatment fluid
Mineral block (containing salt)

Anatomy

GOATS RANGE IN SIZE from about 40cm in the Pigmy breed, to well over 100cm tall at the shoulder of a male Saanen. But they all have long neat heads, bodies that get wider towards the hips, and sloping rumps. Unlike sheep, whose tails hang down, the goat tail is naturally short, and is held upwards or horizontally. With the exception of the Pygmy goat, goat necks are longer than those of their cousin, the sheep, who just grazes grass. With the aid of their long back legs, goats can also reach up to food in bushes and small trees. They are very sure footed. The rough pads on the bottom of the hoof provide them with the "grip" to climb to inaccessible areas to forage. Both sexes, in all breeds of goat, can carry horns. These can grow quite thick and wide on uncastrated bucks. However, some animals carry the gene for natural hornlessness, but these individuals often have fertility problems. In some breeds, both male and female goats can sport a beard, but in others neither sex will have one and purists of the breed will exempt them from registration if they should develop one. Many animals can also carry flaps of skin dangling from either side of the throat area. These "flaps" are filled with cartilage. There appears to be no purpose for this appendage in the present day goat, but it was probably some form of throat guard in its earlier history, like a lion's mane. We call them toggles, tassels, or wattles. In one or two breeds, the purist breeder will not allow these appendages.

Unlike sheep, the goat coat is of hair, not wool, and does not contain the water repellent lanolin that enables sheep to shed water and keep very warm, even in extremely cold weather. Goats do not like the wet, and will quickly seek shelter from the rain. In fact, goats not provided with adequate shelter will succumb very rapidly to chills and pneumonia, and can die very quickly. Goat hair is made up of two layers, a fine, soft under hair, found against the skin, known as "cashmere", which moults out in spring, and a coarser outer hair. In an "entire" Billy, this outer hair is very dense, designed to protect his body when fighting off opponents. After the spring moult, the colour of the coat will often change for the duration of the summer months, but will grow back to its normal colour in the winter. This is particularly true if a white cross-bred goat carries tan patches, or has an apricot colour haze across the body hair.

The Doe, or Nanny goat has a two-compartment udder with large teats, sited between her back legs. If the buck, or Billy is entire (or has been castrated later in life, rather than at birth), he will carry a scrotum, with or without testes. If, however, he was correctly "ringed" at birth, there will not be a scrotum present.

In the mouth, the goat has a set of eight lower front teeth (incisors) and a hard pad opposite these on the top jaw. They also have back teeth (molars) on both the top and bottom jaws that are used for grind-

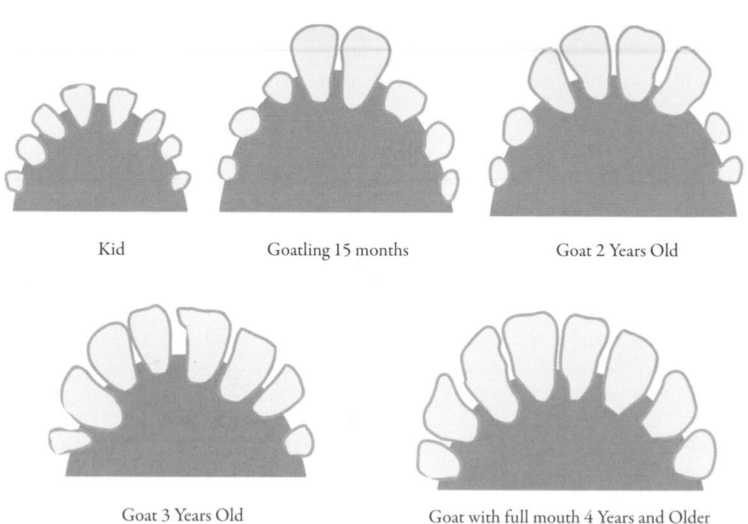

Examining the teeth of a goat will give a good indication of age.

ing. It is possible to age a goat, certainly up to the age of about five years, by examining the front teeth. When it is born, a kid will develop eight "baby" teeth which will be shed two at a time each year for four years. So, if an animal carries four big teeth and four little teeth, it will be two years old.

Each goat foot has two toes. In the domesticated goat, horn can grow quickly on these if the animal is kept on grass, and this will need regular trimming, every four to six weeks.

The goat's body should have good muscle tone without any protruding bones. In aging animals, this muscle will ultimately become quite sunken around the pelvic area and along the spine, though the animal will retain a well-rounded belly area if it is still eating well.

The goat is not only an herbivore, a vegetation eater, but is also a ruminant (like cattle and sheep). This means that the goat has four stomachs. The first is called the rumen, and produces "cud." This is partially digested, and unpleasant smelling food, which is regurgitated back up into the mouth. Freshly consumed, and not too well chewed food goes down into the rumen, where micro-organisms get to work in breaking down the cellulose in the herbage. Small amounts are then pushed back up into the mouth where it is chewed more thoroughly. At this point, the goat is "chewing the cud". This chewing usually takes place when the animal is lying down and is a sign of a contented, healthy animal. It would be wise to familiarise yourself with the position and feel of an active rumen, for if it isn't working correctly you will end up with a sickly animal. Stand behind the goat looking towards the head, and place your hand on the left hand side of the body above the flank. You should feel a sort of rumbling movement; and an ear placed in this area will pick up a grumbling noise. In a sickly animal, vets will listen with a stethoscope for rumen activity.

When the cud is next swallowed, it bypasses the rumen and enters the reticulum, then the omasum, and finally the abomasum, or "true" stomach. Digestion continues throughout this process with the absorption of food into the blood taking place en route, leaving very little waste for disposal. The end product will be large pea sized pellets, which make excellent compost. Any abnormal faeces must be queried. For example, is there a possible worm infestation?

A contented older goat chews the cud.

The Digestive System

esophagus
rumen
kidney
rectum
bladder
diaphram
abomasum
reticulum
udder

Housing

PLEASE BE AWARE THAT IF YOUR ANIMALS are to be housed near human dwellings other than your own, you may need to get Planning Permission to build if stables are not already on the site.

A goat house can be built from any fabric; brick, stone, wood or mixture of all of these, but metal should NOT be used. This can be too cold in the winter months and too hot in the summer. It is particularly important that the roof must be sound and completely leak proof. The goat house should be sited where you will have good access to both water and your storage shed and there should also be a convenient place to deposit the waste that you clear from the stall(s). (This can be added to a compost heap.) The number of goats you wish to keep will determine the size of the housing you need, as each animal will require between 2 and 2.5 square meters. From a hygiene point of view, the bigger the goat house, the better.

We favour the erection of a wall lining inside the house. This adds extra warmth and prevents the walls being damaged, especially if they are built from wood. Sheets of chipboard are ideal, but a rubber membrane like that used for horses will also suffice.

Ready-built goat houses are available, but you will need to build the base (from concrete or slabs) before you can erect these on site. A suitably sized garden shed, with the walls re-enforced inside with lining, would also make an ideal goat house.

Individual stalls inside the goat house, for one or two animals, are our preference.

Individual stalls inside the goat house can be made from chipboard.

A corridor between the stalls and the outside door is very useful.

But goats can be housed communally, provided that care is taken to ensure that each animal gets its share of feed. The dominant animal will always take possession of the hayrack or feed trough, and the others may go short.

If you opt for the stall system and your goathouse is big enough, a corridor between the stall doors and the outside wall and entrance is very useful. This will help to keep out bad weather, and enable you to keep dry while cleaning out the stalls, even when it's raining outside.

Ensure that the floor (preferably, this should be concrete but could also be made from slabs) has good drainage. A slight slope towards the outside entrance or corridor will make cleaning more easy and effective.

We favour a split (stable style) door for the main entrance. This provides good nighttime ventilation in hot, humid weather and allows the animals to view the world on days when they are contained within the goat house. A garden shed door

You will need a good supply of dry sawdust to spread on the goat house floor.

could easily be divided. There should also be good ventilation (say, through the eaves) but no draughts.

Depending on the size of the goat house, there should be at least one window for daylight. This is required for those occasions when animals have to be shut in. It should open wide to allow for maximum ventilation in hot weather. The window should be positioned fairly high to prevent breakage by wandering feet but if this is a problem, a mesh cover should be placed over a glass pane. Alternatively, safety plastic would be ideal. Extra light and an open window is also an asset when cleaning out the house.

Left A custom built, six feet square, goat house by Forsham Cottage Arks.

A larger Forsham Cottage Arks goat house with the favoured type of split (stable style) door.

It is handy to have tethering rings on the outside of the goat house to secure your goat whilst performing tasks like hoof trimming.

It would not be wise to store the food and equipment in the goat house so a secure strongly locked shed must be considered for the storage of hay, straw and feed in bins. This shed could also be equipped with a cupboard for medical and grooming utensils. Keep hay and straw off the ground on pallets to prevent dampness rotting the bales.

HARD STANDING

A solid, preferably concrete, surface (this could also be made from slabs if you are not troubled by moles) outside the goat house will enable you to work in and out of the goat house without slipping about on wet days. It will also provide you with a firm surface on which to groom, trim feet, and generally inspect your animals. An overhang from the roof would also be a great advantage.

If it is not possible for the goat house to be equipped with an electrical supply, you will need to keep a strong-beamed torch to hand for night-time inspection, should the need arise.

Your goats will also need clean, dry, warm beds made from straw or wood shavings. Ideally, this should be a combination of both. Some goats like to be off the ground, so bunk-style shelves would be ideal if space permits. This will mean that your stall dividers, if you have installed these, will have to be higher to prevent the shelves being used as vaults.

If your animals cannot have access to the goat house during the day, you will need to provide a field shelter that will accommodate all of the animals. This should also be provided with bedding. This will be needed in case of rain, strong winds, or excessive heat. Goats need access to shelter at all times. Paving slabs inside and around the entrance area to the field shel ter would be a sensible addition, to prevent ground erosion and mud.

TETHERING

Tethering is not a recommended system of keeping goats; the animals must be monitored continuously throughout the day to ensure that they are not caught up in the tether, and regularly moved onto fresh ground. It is also imperative that the animals have access to shelter and be in constant reach of a full water container. Tethering is a time consuming method and one that causes many injuries to animals entangled in the ropes or chains, as well as sore necks from chafing collars.

Fencing and Security

FOR GOATS, WE HAVE FOUND that the most secure and sturdy kind of fencing is post and rail, with small hole netting (about 3 feet wide) attached under tension to the middle and bottom rails. The top rail of the fence should be at about four feet unless you have particularly leggy animals that like to jump; in which case you will need to go higher. Four feet wide netting is sometimes available but this would require an extra rail (so, a total of four rails between each upright) at about two foot from bottom rail to secure the netting. Obviously, this would entail a lot more expense. The bottom rail will ensure that the Pygmy breed of escapologist cannot get under the fence, while the long legged of the species cannot push through the middle or jump over the top. If you plan to keep only smaller breeds, a well-erected post and stock netting fence would suffice but with a tension cable applied along the top to prevent crushing. Well-kept goats should not want to escape, but it is better to be prepared for the odd rebel who fancies your neighbour's vegetation. Goats easily push around weak fences, and love to rub themselves along them, especially in the spring and summer when they are shedding their winter coats. Any potential escape routes will soon be spotted. You should therefore provide the most secure fencing that you can afford, with possible future problems in mind.

Post and rail fencing with wire netting is essential as goats are notorious for escaping.

Never use barbed wire. This is an accident waiting to happen when the wire comes into contact with a curious animal like the goat, and faces and necks can get badly torn.

Some goat keepers use electric fencing in conjunction with other less secure regular fences, but this is not a system that we

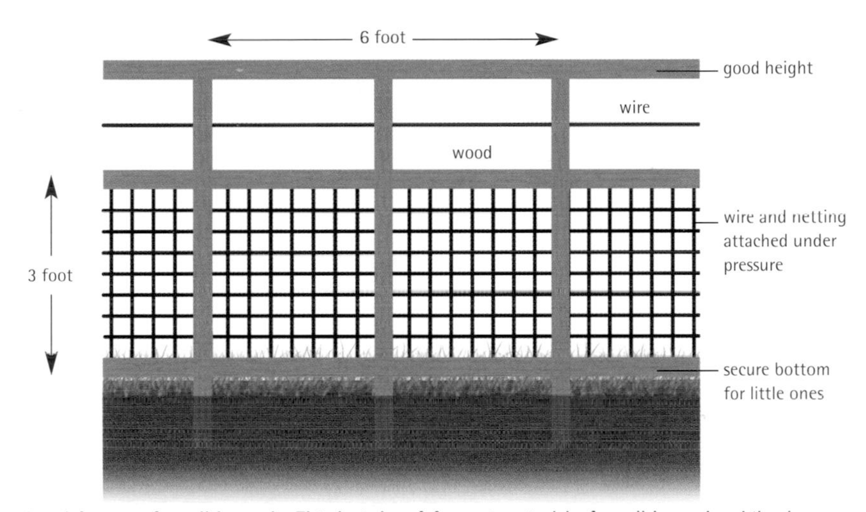

Post and rail fencing for all breeds: This height of fence is suitable for all breeds while the lower rail to prevent tiny kids from escaping

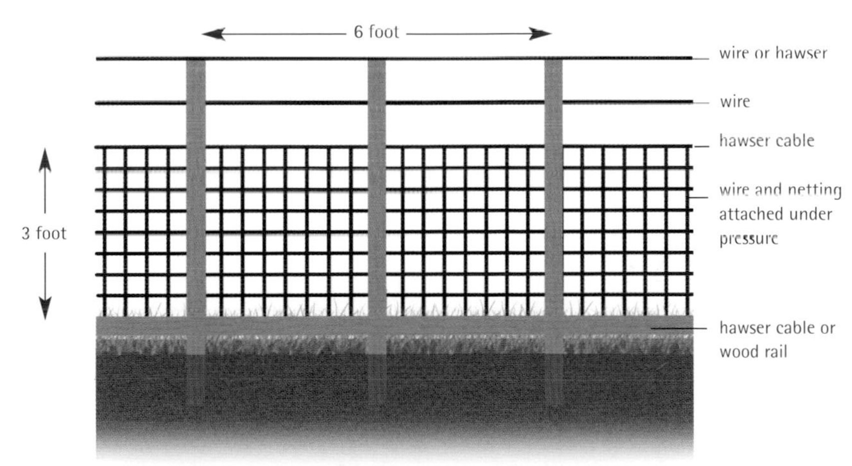

Stock fence suitable for Pygmy or smaller breeds

Goats love to eat tree bark so erect fencing around your trees too.

would recommend, particularly with horned animals. The disadvantages of electric fencing are significant. There is the worry that the battery might fail when you are not around and the animals get out or get caught up. We have also seen horses and sheep entangled in electric fencing which is most distressing for all concerned. There is also an environmental concern, plastic strips do not look very attractive and will inevitably tear and get blown away, littering your surroundings.

If you have two or three small fields, a rotational system of grazing would ensure regular clean pasture and allow vegetation to grow. Alternatively, a large paddock could be divided into smaller areas depending upon the size and number of animals that you have. They could spend ten days in one section then go on to the next. The disadvantage of this system is the access to the goat house. A corridor might be the solution.

SECURITY

Ensuring that your animals are contained within your property is paramount to good relationships with your neighbours. Goats can be destructive and will avidly strip bark from trees and munch vegetation; this is not just bad for your neighbour's property but could also be detrimental to the animals, as they may poison themselves. Also, if your goats get onto a road and cause an accident, you will be liable. You may want to consider taking out insurance for such events.

Not only must your animals be secure, but they must also be safe. Their environment must not harbour any harmful substances such as rubbish or polythene. Do not dump your garden waste in their compound unless you are 100% sure it is safe, as it may easily contain plant matter that is either harmful or poisonous. Curiosity can kill. You should also check for any projections such as nails or damaged fencing that could injure or trap a goat.

Food and Water

WATER

IF YOU ARE FORTUNATE ENOUGH to have a standpipe on your property, and decide to erect your goat house nearby, you will need to check for the location of the pipes, lag the standpipe, and use a manageable trough for providing safe outside water. This will require periodic cleaning out as algae builds up around the sides for providing outside water. If you don't have a standpipe, you will need to invest in a good hosepipe. This can be attached to the nearest water source, reeled in or out, and stored to prevent damage from the animals and freezing. These preparations will save you a lot of carrying of buckets backwards and forwards across insecure ground in bad weather. Fresh water must be available at all times.

FOOD

A good coarse mix goat feed should be available from agricultural merchants, but if you decide to keep the Pygmy breed of goat, you might want to use a special Pigmy blend of mix, which isn't so high in protein. This will avoid the problem of fatness to which Pygmies are prone. If this mix is not readily obtainable, then you will need to reduce the amount of the regular

A good supply of clean water should be available in the paddock. Note the insulation to avoid the tap freezing in winter.

Mount a salt lick somewhere handy so your goats can serve themselves.

coarse mix you feed. Unless you are breeding or milking, you should avoid the dairy blends. We feed a mixture of goat mix with Alpha, a product that is used for horses. This provides bulk satisfaction without the risk of too much protein, which can cause health problems.

Good quality hay should be fed ad lib, particularly in situations where there is little browsing material. Unless they are desperate, goats will take only the tops of long grass, weeds, and leaves, (goats love the autumn), so a constant supply of roughage is necessary. Do not be tempted to feed extra mix, but use it as a supplement, feeding a little more to older animals in the evening. One meal of mix a day should be adequate for a young animal, but you could split the daily quota into two small meals, particularly with the Pygmy breed. Offering food in the evening is also a good way of training your animals to come in at night if they are being naughty, which they often will be.

In the winter months, a small amount of Lucerne and grass nuts is a useful supplement when the field forage has lost a lot of its nutrients. This can be fed in the evening when the animals are put away at dusk. You should also ensure that the goats also have hay, remembering that the

Good quality hay should be available at all times to supplement grazing material.

winter nights are longer.

If and when, you decide to start milking, the food supplies will need to be slightly different as the lactating female will need food for maintenance, processing and the milk. Many people have vary-

Foods for your goat

Leftover cabbage, lettuce, tomatoes, and apples can be made into a delicious salad for your goat.

Dried sugar beet flakes

Goat mix

Alfalfa grass blended with hay, straw and molasses

Crushed oats

A good supply of clean bowls for eating and drinking is essential.

ing opinions as to what to feed so talk to folk who milk their animals and be guided by them.

Pregnant females will require extra rations on an increasing scale a few weeks before they kid, as this will be the time when the embryos grow the most rapidly and will drain the mother of her body stores if she does not get additional food. After kidding, the extra feed should be given on a decreasing level. Not only does "mum" need to get back in condition, but she will also be supplying milk to her kid(s). She may well have produced two, three, or even four offspring. Again, it would be wise to talk to experienced breeders for advice.

Your local greengrocer may well be a source of unwanted fruit and vegetables particularly in the winter; and maybe an independent bakery will sometimes have left over loaves. Your home vegetable waste, cabbage leaves, lettuce, tomatoes, apples etc stale bread, will all be very welcome little treats. They also love cream crackers, but no dairy or meat products must be fed to goats. Too much brassica feed for a milking goat will also affect the taste of milk, but it is not poisonous.

The following plants are poisonous in varying degrees and care should be taken to ensure that they do not grow on any land to which the goats have access:

Poisonous Plants

ALDER
ARUM, cuckoo pint (*Arum maculatum*)
AUTUMN CROCUS (*Colchicum autumnale*)
BRACKEN (*Pteridium aquilinum*)
BRYONY Black and white (*Tamus communis & Bryonia dioica*)
BUTTERCUP (*Ranunculus*)
CELANDINE (*Ranunculus ficaria and Chelidonium majus*)
DOGS MERCURY (*Mercuralis perennis*)
FOXGLOVE (*Digitalis purpurea*)
Green stuff from flowers including delphiniums , hellebores or any bulbous plants such as daffodils or tulips
HEMLOCK (*Conium maculatum*)
HONEYSUCKLE (*Lonicera*)
HORSETAILS (*Equisetumspp*) **particularly in hay when dry**
IVY (*Hedera helix*) **berries**
LABURNUM (*Laburnum anagryoides*) **foliage and pods**
LAUREL and other evergreen shrubs
MAYWEED (*Matricaria maritima*)
NIGHTSHADES (*Atropa and Solanum*)
OAK (*Quercus*) **acorns**
POTATO (*Solanum tuberosum*) **haulm and green tubers**
PRIVET (*Ligustrum vulgare*)
PRUNUS the withered leaves of plum, cherry etc.
RAGWORT (*Senecio jacobaea*) **particularly when dried in hay**
RHODODENDRON (*Rhododendron ponticum*)
RHUBARB (*Rheum raponticum*) **leaves**
SPINDLE TREE (*Euonymus europaeus*)
THORN APPLE (*Datura stranonium*)
TOMATO (*Lycopersicum esculentum*) **foliage**
WATER DROPWORT (*Oenanthe crocata*)
YEW (*Taxus bacata*) **berries and foliage**

Ragwort

Choosing your Goats

GOATS ARE HERD ANIMALS and the new goat keeper must consider this when deciding on the number of animals to be kept. We always recommend a minimum of three animals, but a pair is acceptable. However, one must consider the possibility that if one of the two dies the owner is left with a grieving animal and a replacement must be sought as soon as possible. With three animals, there would be more time to replenish the numbers. Once the number and breed has been decided upon, and all necessary arrangements have been erected and installed, you can seek out local breeders via the British Goat Society, or you might want to contact local animal sanctuaries for goats that need re-homing.

For the beginner, the Pygmy is ideal from a size point of view but Toggenburgs or Golden Guernseys would also be suitable. Although they are larger animals, they have good temperaments. You should

Very young goats may look attractive and cute but can be a source of mischief.

always view your animals before making a final decision, checking them for sound appearance and good health. Your herd could be all female, all wethers (castrated males) or a mix of both, with two girls and one boy to make a trio.

Older animals can also be ideal for novice keepers, as they will already have been handled, hopefully kindly. It will also give the goat keeper a sense of well being from being able to give these animals a great deal of attention in their latter years. Very young animals may look cute and attractive but they can be a source of mischief and hard to handle. You must also consider that, on average, goats will live to about 12 to 14 years. Will you be able to cope with them for that period of time? If you are purchasing your animals for the benefit and education of your children, you should also consider that they may lose interest in the goats, so that the parents are left with total responsibility for the animals.

Once you have decided on your goats, you will then need to address the subject

Goats love cream crackers!

New goats should be left in peace to take in their new surroundings.

of transport. For larger animals, you will need to buy, borrow, or hire a trailer or horsebox that has easy access for loading and unloading. Very small kids and goats can be transported in large dog cages.

THE DAY OF ARRIVAL

At the time of writing, a UK-based goat keeper will require a movement licence to transport their newly acquired animals. This is obtainable from Trading Standards or (in some cases) the person from whom you obtain the animals. You will have to enter the holding number of your land on the licence.

Ensure that your paperwork is available for the chosen day, and that your route is planned to ensure the easiest and least stressful journey for the animals. DEFRA (the Department for the Environment, Food and Rural Affairs) regulates the transportation of animals, particularly for long journeys, and you must consider the welfare of the animals at all times. It might be a good idea to take collars and leads with you, and you may find a packet of cream crackers helpful.

Before you leave home to collect you goats, you should make sure that the goat house is ready to receive its occupants. Food and water should be left in place so that on their arrival, the animals can be unloaded and left in peace to relax and take in their new surroundings. It will take three or four days for the goats to settle. During this time, you should leave them alone, preferably in the house, and move around them in a slow, quiet manner with no sudden movements or noise. This will help to win their trust while you are feed-

It is good to share the tasks of goat keeping with the children.

ing them or filling their water buckets. Once they have checked out the grazing area, your new acquisitions should find their own way back to the stable without any physical help from you, but a tempting bowl of treats will offer encouragement until they feel completely relaxed and at home.

Routine Maintenance

Make sure you fill up the outside hayracks daily.

DAILY, WEEKLY AND MONTHLY ROUTINE

DOMESTIC ANIMALS do well on routine and goats are no exception to this rule. They are creatures of habit and feel very secure and at ease when the day follows a regular pattern. Their daily requirements must fit comfortably into your life for 365 days a year, with no days off and no over sleeps, so you will need to work out a routine that suits you. Otherwise, looking after the goats will become a chore, and you will resent each day that you have to attend to their needs. Before you buy your pets, you should make out a plan that you think will work for you. If you are a family, then share the tasks, particularly at the weekend when the children are at home. Your goats will be with you for a long time, and they are not easily re-homed.

SUGGESTIONS for your day

- You should pick a time in the morning to visit the goats that will vary little throughout the year (depending on the hours of daylight), and that fits in with your family life. Goats get very noisy when they are hungry or need attention, so it is best to get them sorted as soon as possible.

- You should first feed the concentrate meal and fill up the outside hayracks if the weather permits outside activity for the animals. If the weather is too wet and cold

'Plump up the cushions' by replenishing you goat's bedding straw daily.

for the goats to go out, the animals will probably need to stay indoors, so you will have to work around them.

- Check that your animals have eaten and look well before you release them to go outside. This would be the ideal time at which to administer any medication, and if it is time, to worm them or carry out feet trimming. In springtime, you could start your grooming once you see hair beginning to loosen up and fall away.

- Once the animals are out of the house, remove the water buckets and clean out the soiled bedding, retaining as much as possible of the dry. "Plump up the cushions" by forking over the litter then replenish the top with some new straw; any hay left over in the racks can be added to the bedding to help cut costs. "Waste not, want not". If the weather has been really good, you will find that the beds are quite dry and need little attention.

- In the winter months, some folk advocate a weekly or fortnightly clean keeping the bedding in situ and adding fresh to the top. The number of animals kept and the state of the weather must determine the suitability of this system. If it is very wet under hoof, it really isn't an option as the animals will bring in muddy feet and sometimes wet bodies and contaminate their bedding. I prefer a daily clean; it isn't as back breaking as moving a week's supply of manure. If the house smells strongly of ammonia, you should clean daily.

- In the summer months, leaving the cleaning-out for a week will also attract flies, which goats really hate, and will encourage lice and mites. This is also a good time to have a really good clean all around

Goats will find their own way back to the gate at bedtime.

the racks and window frames and to disinfect with a weak solution of Jeyes fluid. You should also clean the water buckets if they are soiled, then refill and return them. Clear up around the goat house, ensuring that there is nothing left about to endanger your animals, such as baler twine, polythene bags, or syringes from worming etc.

- In the evening, at whatever time dusk falls, your animals will find their way to their stalls ready for bed or, if field bound, they will wait at the gate, so be ready to let them in. Bedtime will vary throughout the year as the seasons dictate longer or shorter days, but it will be down to you to shut them safely indoors. In wintertime, you should give the goats a small overnight feed of hay. Goats are quickly trained to follow this bedtime routine, with a little encouragement of meal for a few days after they first arrive.

Grooming

GROOMING IS AN IDEAL ACTIVITY for you and your goats to really enjoy each other's company in a calm and soothing way. It is the best of all contact times, providing moments when you can tell them how beautiful or handsome they are and how life was dull before their arrival. Goats love to be brushed, but need assurance as to your intentions when first you start, so once they have settled into their new life after, say two weeks, commence the ritual for just a few minutes to introduce them to the experience. Use a collar and lead connected up to a secure hook or post, and commence with gentle strokes around the neck and shoulders. Avoid touching the horns as many goats do not like the sensation and will get very fidgety. Gradually work your way along the back and down the legs. This is also a sensitive area, so be prepared for a lot of movement.

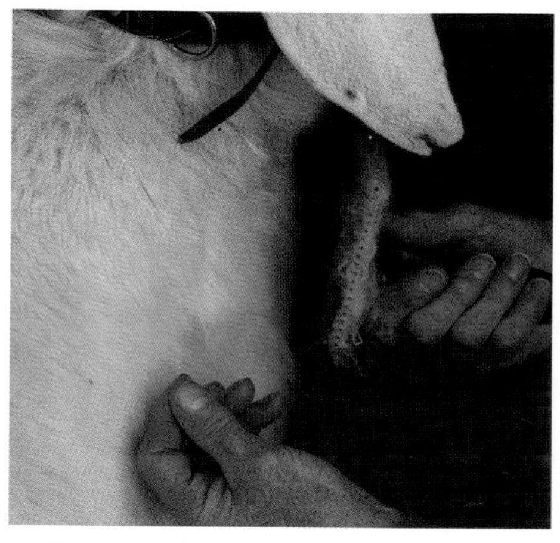

Groom weekly with a rake-type comb to remove moulting 'under' hair.

A light weekly groom should keep your animals' coat looking good. But daily attention may be beneficial in the spring, as the under hair will be moulting, and the coat may be looking really scruffy. Very expensive grooming equipment is available, but good quality dog grooming utensils would be quite adequate, including a rake-type comb and a wire dog brush to remove loose hair, together with a dandy brush to smooth the surface. Long coated breeds will always need the rake on their breech hair to remove tangles.

In the winter, you can groom just once a week with the dandy brush (without using the rake). But do not be over zealous and pull out that warm, very much needed,

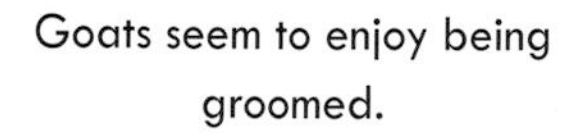

Goats seem to enjoy being groomed.

A weekly grooming with a dandy brush will maintain your goat's coat in good condition.

under hair. A firm smooth motion to stimulate circulation and polish up the hair is all that's required.

A treat of some kind would help to make the experience a pleasurable one for your goats.

HOOF TRIMMING

Hoof trimming is a very important factor in the welfare of your goats and will need to be addressed every 4 to 6 weeks. If it is done regularly the foot will keep its shape and lameness with be prevented. If you

Three types of commercially available hoof trimmers.

know of a goat keeper or shepherd who can show you how to carry out this procedure then it will become quite an easy operation. Most local goat clubs will arrange for demonstrations. Run your hand from the shoulder or rump down the leg as a warning to the goat that you are going to lift the foot; do not just grab at the foot and startle the animal. Either place the leg between your knees as you face towards the back of the body, or to push your body against the body of the animal and support the goat's leg with your inside leg. Using the tip of the foot clippers, remove any mud from the base of the hoof and between the toes (a piece of rag can be pulled through the toes to clean back to the skin). You should then cut the horn in stages, back to a level with the pad in the centre to the white line. The very ends of the toes should also be snipped back to avoid curling. When you have removed both sides of each toe you must then level the heel flap so that the whole foot is flat. If any side of horn bulges outwards with mud stuck up inside it, you must cut away the horn and remove any debris, as this will be uncomfortable and harbour bacteria. Sometimes the "quick", or cuticle, can be accidentally snipped and will bleed. Do not panic, but when the trimming procedure is complete, spray the foot with your antiseptic spray. If the bleeding is profuse, a small tuft of hair or cotton wool can be stuck on to the wound. You should then leave the animal standing on the hard surface for a few minutes. The goat may limp for a day or two but a daily spray should ensure that the cuticle doesn't become infected. Blood always makes an injury look a lot worse than it is in reality.

When the trimming is complete, check between the toes for any soreness (a condition known as SCALD can develop here). If this is present, spray between the toes with your antiseptic. If the goat has been limping, and a foot or feet smell rotten, then a condition called Foot Rot is present. Treat this as suggested in the Health section.

WORMING /DRENCHING

This is actually the messiest of jobs, although the instructions and the photographs shown in books make it look easy. You will probably end up covered in the wormer suspension or whatever it is that

Hoof Trimming

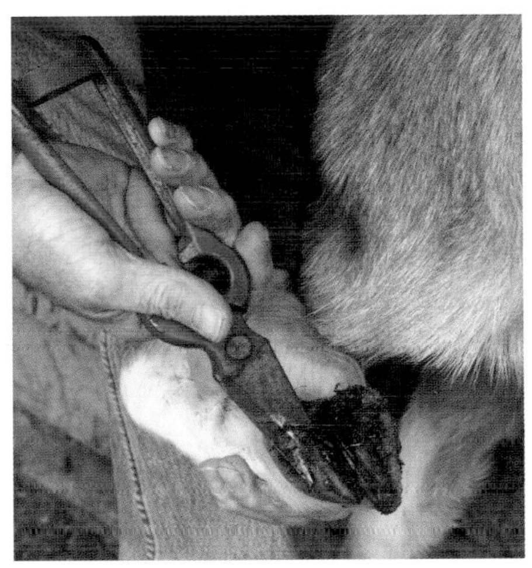

1 Grasping the goat's foot securely trim the outside part of the hoof first

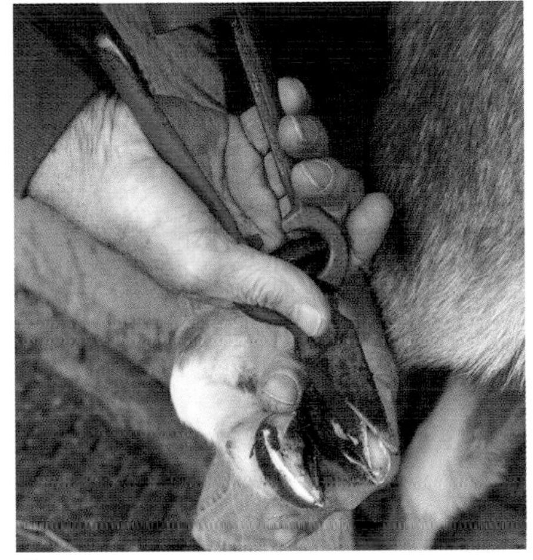

2 Trim the inside part of the cloven hoof secondly.

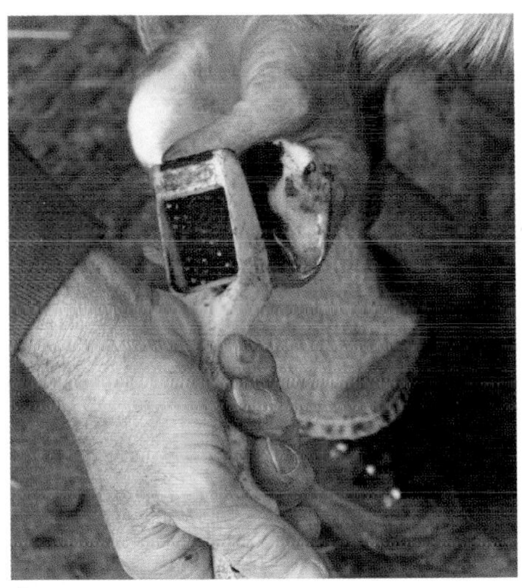

3 File the hoof to remove any rough edges.

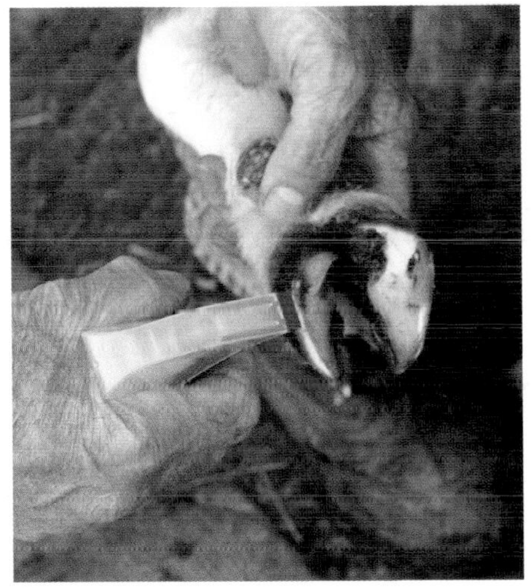

4 Spray with anti-bacterial fluid to avoid infection.

you are trying to get the goat to swallow. There are some injectable wormers, but drenching is the best way to ensure good results. Powders and pills are available but goats are clever enough to eat around powders sprinkled in food or to retain a pill in their mouths and expel it later.

Try to establish an approximate weight for each animal and read the recommended dosage on the wormer container. Tip a small amount into a cup or container so that you can draw up the solution into the syringe (syringes of about 10 mls or 20 mls in size are ideal. New ones would be available from your agricultural merchant or vet.)

THE ONE MAN ATTEMPT

Straddle the animal with your knees around the neck. Place your left hand under left side of the jaw, lifting the head into a straight position. Then place your left thumb in the gap between the front and back teeth, holding down the tongue, to pry open the jaws slightly. On the right side of the mouth, place the syringe or bottleneck as far back on the tongue as possible, and gradually pour in the drench, slowly enough to enable the animal to swallow. There should be no sudden squirts, or fast pouring in case the solution goes into the windpipe and down into the lungs. You will usually lose some solution on your hands or on the animal's mouth, jaws and neck. If the goat is very small, a kneeling position with your body pushing against the animal's side might achieve results. In time, you will develop your own technique, hopefully without too much stress to either party.

TWO MAN ATTEMPT

One person straddles the animal as previously mentioned or pushes the goat's side against a wall, and uses both hands to hold open the jaws. The second person slowly feeds in the fluid. Two of you get messy.

Once the operation is accomplished, a cracker will renew the animal's love for you.

If you are grazing only one paddock, regular worming every 5-6 weeks between March and the end of October should ensure good control. But if a rotation system is in use, then worming could be carried out only two or three times a year. If in doubt, talk to your veterinarian about how often to worm, and about which products to use.

Always store your drenching solutions in a safe, childproof cupboard along with any other preparations that you may be keeping for your animals.

BEHAVIOUR

Generally, goats are affectionate, people-loving creatures that love attention and will willingly follow you around and rub against you if given the opportunity; a scratch would always be welcome. They are intelligent and very quickly learn your routine, where they sleep, and that "goodies" often arrive in polythene bags.

When interacting with fellow goats they will often head butt, mount each other, and nibble each other's coat; this is all quite normal. One goat will always strive to be the Alpha animal and demonstrate his or her superior strength. The others will fall into the order accordingly (the pecking order). Do not be tempted to play "butt" with a goat; this will encourage the animal to butt if it wants attention; not a good habit particularly if children are involved.

In the mating season, an entire (uncastrated) male will become very possessive of the females and should be avoided. Equally, a nanny or doe that is in season may sometimes go off her food and become more affectionate towards you, this is all quite normal.

Goats often play-butt each other in fun.

Goats are affectionate creatures that will follow you around.

Leading your Goat

It can be very useful to lead your goat by a collar and lead, just as you would walk your dog. You may, for example, find it helpful to tether the animal while you carry out maintenance work, administer medication, or lead it into a trailer. To make the experience as stress free as possible, you need to familiarise the goat with the equipment on a daily or weekly basis. As goats are inquisitive, climbing animals it is highly inadvisable to leave their collars on permanently, which could easily become tangled in bushes, or caught around a horn during fighting games. Collar wearing can also lead to a goat's neck hair becoming spoilt and thin.

On the whole, goats are no less amenable to being led than most dogs, and the training procedure is much the same. Introduce the goat to the idea of wearing a collar at an early age, at times when you are supervising the animal. At first, goats may prove reluctant to lose their liberty so keep the collar and led behind your back. Apply the collar firmly, but without over tightening it.

When your goats have got used to the collar, you can then attach the lead, and start to take them for short walks around your property. If you are right handed, hold the loop end of the lead in your left hand, and place your right hand at the collar end, touching the collar itself. This means that if the animal moves unexpectedly, you can control it with both hands, rather than being pulled along by one. If you are left handed, you will probably find the opposite position more comfortable and secure.

Goats soon get used to "walking" in this way, especially if good behaviour is rewarded with a cream cracker or carrot. You need to lead the animal forward by walking at its side.

Even quite strong goats can be trained to walk with children and young adults.

When you feel more confident, it can be very enjoyable to take your goats for longer walks away from home, giving them the opportunity to browse in the hedges and verges. But the inevitable danger from passing traffic means that you must have complete control of your animal at all times.

Health

UNDERSTANDING WHAT IS HEALTHY in your goats gives you a kick start to knowing when an animal is unwell; daily observation and examination right from arrival will help to quickly determine if there is a problem.

Indications of a healthy goat are as follows:

The animal is eating well, and is chewing the cud.
He is being sociable, is alert, lively, and jumps without hindrance.
He has a healthy looking coat, that is not wrinkled or pinched, but smooth.
He has clear eyes, with no unpleasant discharge.
He has a cool dry nose without discharge.
His faeces are normal firm pellets - not large clumps or "cow pats".
His urine is clear and straw coloured.
His general body condition looks well nourished, but not bursting

It is not common for a goat to be severely ill but it does happen from time to time, so it is necessary for every goat keeper to be prepared to nurse a sick animal. There are many ailments from which a goat might suffer, but I shall deal with only the more common problems to avoid frightening the beginner. If you are at all worried about an animal, you should contact your friendly goat keeper or breeder and be advised by them as to whether the veterinarian should be called. You should also keep a basic first aid kit.

MEDICAL CUPBOARD CONTENTS / FIRST AID KIT

Antiseptic lotion e.g. Dettol or Savlon
Antiseptic cream or spray
Bicarbonate of soda
Clinical thermometer - preferably not a glass one
Cotton buds
Cotton wool
Crepe bandage (5cm/2 inch)
Drenching bottle (not a glass one, a plastic water bottle is ideal)
Epsom salts
Gauze or bandage (5cm/2 inch)
Glucose
Liquid paraffin
Scissors (round ended)
Syringes for worming. These can be re-used after sterilization with sterilizing fluid.
Table salt
Terramycin antiseptic spray or foot rot spray (often the same product)
Tea bags
Vegetable oil
Worming formula *
An old blanket and baler twine for a makeshift coat
Pieces of old cloth for cleaning feet

The most common ailments of a goat will be to do with the stomach, the feet, the skin, and (possibly) the eyes.

The normal temperature for a goat is between 102.5 and 103 degrees Fahrenheit, or 39 degrees Centigrade. If the goat's temperature is half a degree either way, do not panic. But check it again in half an hour. If it has risen or dropped and you have a friendly goat keeper contact, call them. If not, call your vet with a note of all the symptoms presented. But if your goat is spilling the cud, or has breathing problems, call your vet immediately.

If an animal is collapsed, it should be propped up between two bales of straw. Warmth is essential so tie a piece of old blanket around the body. Keep the animal away from draughts and try getting it standing and moving around. Contact your vet immediately.

Good hygene is important when bottle feeding.

SCOURING (diarrhoea)

This problem is very obvious and will probably be the result of a sudden change of diet, an internal parasite infestation, the ingestion of poisonous plants or substances, or an internal infection.

Ask yourself if the animal has had a change of food and when was the animal last wormed. If this was longer than ten days, worm the animal again and wait 24 hours UNLESS:

1 If there is blood in the faeces or rapid deterioration, you should contact your veterinarian immediately.

2 If the animal shows signs of pain (crying out, lying awkwardly, fidgeting) if is spilling the cud (vomiting), or has difficulty breathing, you should suspect poisoning and call your vet. You should also try to establish what might be the source of the poison.

Scouring can also occur in a bottle-feeding kid. Replacing the milk with warm water and glucose for about 24 hours should clear the problem, but if it persists, you should suspect worms and drench accordingly. If scouring continues, consult your veterinarian before the animal becomes dehydrated. If you suspect dehydration, pinch a section of loose skin. If it stays "pinched" administer a hypotonic electrolyte solution to begin rehydration while you await your vet. He will probably recommend that this be continued until the diarrhoea clears. Sachets of electrolyte solutions are available from the vet

Observe and inspect your animals daily for signs of ill health.

but a homemade solution will suffice if necessary. To make an electrolyte solution, mix together:

1 litre (2 pints) warm water
½ teaspoonful table salt
¼ teaspoonful bicarbonate of soda
2 tablespoonfuls of glucose or honey

Drench the solution throughout the day, using a plastic bottle, or as the vet directs.

LAMENESS

Lameness is present when an animal begins to limp, is reluctant to stand or move around, or kneels when grazing. You should catch and contain it, and begin a thorough check. Begin with the feet and work up to the shoulder.

The most common cause of lameness will be dry compacted mud, grit, twigs etc. between or under the toes. Clean this out with your fingers and some cloth. If the skin is damaged, spray the wound with some antiseptic. If thorns or stones have punctured the foot, clean it off and spray the wound with antiseptic, but keep a close watch for swelling. This would indicate the presence of infection, or a possible abscess. If this occurs, contact your vet.

The infected matter will need to be expelled by lancing, and antibiotics given. It will then be the responsibility of the goat keeper to clean out the wound every day with hot salt water or an Epsom salts solution until it is clear of infection.

SCALD

Scald is an infectious bacterial disease that occurs between the digits (toes). The skin looks sore, and is sometimes swollen with a grey discharge. The foot should be sprayed with a Terramycin spray and treated daily until the infection has cleared up.

FOOT ROT

Foot rot is a highly infectious bacterial disease of the feet that invades the underlying tissue of the sole of the foot. It gives off a foul odour so you will often smell it before you see it. "Rot" causes severe lameness and if it is not treated, the infection will eventually attack the bone as gangrene. Cut away as much dead hoof as you can, particularly at the side where there may be a gap between the hoof and foot. Do not panic if the foot bleeds. Then soak the foot in a solution of formalin if this is available (1 part to 9 of water). If not, strong salt water or antiseptic water can be used. You should then spray with a Terramycin foot rot spray. Clean up and dispose of all hoof parings and disinfect your cutters and the ground on which you have treated the goat. Treat the infection daily until it has cleared.

If any foot disorders persist or the animal presents with additional symptoms, contact your vet.

Any swellings or bruising in a leg may indicate dislocation, breakage or sprain so contact the vet.

MASTITIS (udder infection)

Mastitus is caused by a bacterial infection, which usually enters the udder via the teat canal or in some cases through injuries to the skin. The infection can also be caused by bad milking techniques. This can affect one or more divisions (or quarters) of the udder. Lameness in the back legs may indicate an infection in the udder, so if nothing is found in foot or leg as a cause, check out the udder. If it feels hot or hard, or if it is discoloured or looks painful, or is swollen and enlarged, you should suspect mastitis and contact your vet. Untreated, this infection will eventually kill your goat as it spreads through the blood stream.

If the goat makes a full recovery, but the udder does not the infected quarter may slough off. Nannies that have recovered from mastitus may not be able to rear further offspring and should not be bred from.

INTERNAL PARASITES (worms)

A small amount of roundworms are always present in the gut of an adult goat. Young goats cannot cope with this as they have built up no immunity and so should be checked and treated more frequently

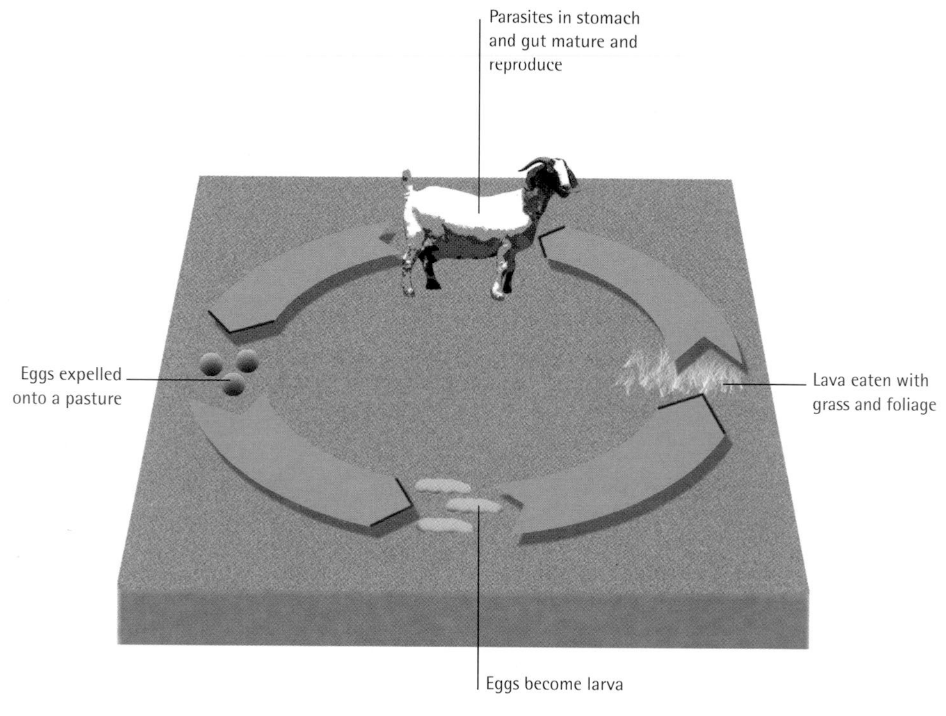

A Cycle of Parasitical Infestation

than the adults. If the animals are grazing on infected pasture the balance will be upset and diarrhoea will occur with an eventual loss of body condition, dehydration, and possible anaemia. If this is left untreated, your goat will die. A good worming programme or grazing rotation will help to prevent a large infestation.

Just like cattle and sheep, goats can become infected by liver flukes when grazing on low, wet pasture. These are tiny fresh water snails that, if ingested and left untreated, will cause permanent damage to the goat's liver. Some worming preparations also contain an anti-fluke compound. Good land drainage will help to alleviate the problem but the application of a molluscicide (a snail killing chemical) is not recommended as it damages the environment and kills songbirds. You should seek advice.

Although goats housed in a yard without access to pasture are less likely to become infected by internal parasites, they can still pick them up from foliage you bring in. In other words, it would still be prudent to adopt an anti-worm regime but this could be done less frequently. Goats browsing on shrubs and trees are also less likely to pick up large quantities of parasites but must still have some preventative treatment.

There are three different compounds for worm management, and it is recommended that your regime includes all three on a rotational basis. Each compound attacks different worms and ensures that the parasites do not become resistant. It would be a good idea to talk with your veterinarian about internal parasites and set up a prevention regime.

EXTERNAL PARASITES AND SKIN DISORDERS

It is quite common for goats to be infested with lice or mites (mange). They are often seen crawling through the hair, and some mites will bury themselves in the skin causing the goat to scratch or nibble at the infected area. Preparations are available to treat the obvious external species but a vet must be consulted if the skin mite is suspected.

FLY STRIKE

Fly strike is the most distressing of skin complaints. This is when a fly is attracted by a wound, foot rot, or a dirty tail area (particularly in muggy warm weather), and lays eggs on the goat. Within 24 hours, the eggs hatch to maggots that work their way down to the skin and chew into the skin and flesh. Untreated the animal will suffer enormous discomfort and die of stress. Preparations to repel flies are available to spray on the coat but your vet must recommend these, as some are too abrasive for goatskin. Treat any wounds with antiseptic and keep the coat free of faeces and urine.

Keep your goat's coat free of any soiling that could attract flies.

RINGWORM

Ringworm is not very common in goats. This is a fungal disease that causes bald patches or scabby lesions; it is infectious to humans so avoid contact and consult your vet immediately.

If an animal is not eating, or is lethargic, unsociable, or standing in a corner away from the other animals, take the animal's temperature via the rectum.

A digital thermometer is the easiest and safest to use. Apply a little Vaseline to the instrument and gently place at the entrance to the rectum twisting slightly as you push it in for a short way. Hold the thermometer in place for a minute, remove it and take a reading. You should clean and disinfect the thermometer before you put it away.

ORF (Contagious Pustular Dermatitis)

This is a viral infection usually found in sheep but it can also affect goats and humans. Pustules full of infected matter appear on the lips and gums. An infected suckling kid will infect the udder of the nanny. The infection can be treated by spraying the infected area with a suitable antibiotic, although the infection should eventually clear up unassisted. You should always wear gloves or avoid skin contact, as this disease is transmittable to humans. If the condition deteriorates, or the pustules invade the mouth or affect any other body areas, contact your veterinarian. A secondary infection may be present.

CONTAGIOUS OPHTHALMIA

This is a disease of the eyes. In sheep, it is know as Pink Eye. A similar illness found in cattle is known as New Forest Eye. It is a form of conjunctivitis, which if it is left untreated will cause ulceration and sometimes even permanent blindness. The organism rickettsia conjunctivae is present in some normal-looking "carrier" animals. It can become active and multiply following damage caused by extreme brightness (reflection from sun or snow), or the presence of a foreign body (flies, dust, or anything that traumatises the eye ball). Symptoms will include a discharge, redness of the eyeball, or (in an advanced stage), cloudiness. If any of these symptoms are observed, contact your veterinarian, who will prescribe either an ointment or an antibiotic injection to the lower eyelid.

EUTHANASIA

A veterinarian or a licensed slaughterman must carry out any euthanasia of sick or aged goats. In the UK, the carcass must then be disposed of via a licensed company who will take the body to a special incinerator for farm animals. Your friendly goat contact may be able to recommend someone but if not, the Animal Health Department of your local Trading Standards will be able do so. Animals that die naturally must be disposed of in the same manner.

NOTIFIABLE DISEASES.

It is unlikely that the goatkeeper will be confronted with the following diseases but they must be aware of the law relating to them. Further details will be available from DEFRA in the UK or the equivalent Animal Health department in other countries.

Amongst the diseases involved in the UK are Scrapie in sheep, (the equivalent has not yet identified in goats), BSE in cattle, Foot and Mouth disease, and Anthrax. None of these are common in the UK, but they are all acutely fatal.

Breeding

THE AMATEUR GOATKEEPER should not consider breeding goats, as a great deal of experience in goat keeping is essential to avoid suffering and the loss of animals. A great deal of expense will also be involved. However to further the knowledge of the beginner and explain the reasons for unusual goat behaviour at particular times of the year, I shall briefly discuss the sex life cycle of the goat.

As already mentioned, the uncastrated male goat will have an unpleasant odour, which will become stronger during the mating season. He will spray his face and body with urine, which will carry traces of testosterone, to make him more attractive to a nanny. This is a rather overpowering aftershave. He will also raise his head and roll his top lip to taste the air. He may also become quite aggressive, which is a good reason not to keep an entire buck. He will also mount a doe that comes into season. It is not really necessary for amateur goat keepers to keep uncastrated males, as even if they decide to put a doe into kid (when they have gained several years of experience), stud goats are available from experienced breeders.

A healthy goat will follow its instinct to breed.

In temperate climates, the female goat will come into season (heat, or oestrus) and be ready to accept male service every twenty-one days. This cycle is triggered by the shortening days of late autumn and early winter. This phase will continue into spring, up to late February or March. Usually, the doe is receptive for just one day. Indications that the female goat is in season may include a lot of tail wagging, an almost clear discharge from the reddened and swollen vulva, continuous bleating, and loss of interest in food. She may have some or all of these symptoms. Before obtaining female goats, it would be wise to find out if there are any entire males being kept in the neighbourhood as their presence might excite the females into trying to escape. He would try to do the same if he became aware of receptive does in his part of the world like an un-neutered dog seeking a bitch.

Wethers are castrated males who make excellent, very loving pets particularly when they have been castrated correctly at birth; they may exhibit some of the entire male behaviour at season time, tasting the air and mounting other animals, but this is quite normal. If however, a supposed wether has a strong odour, is irritable, and behaves more like an entire Billy, then you should suspect that one or both testicles have been retained in the body cavity. This situation may have been created by incorrect castration by an inexperienced goat-keeper; and will have caused the poor creature a lot of discomfort. Contact your vet. If it is a matter of retained testes, they will need to be surgically removed under anaesthetic.

Castrating is done a few days after birth, and should only be undertaken by an expert. Castrating the older animal is a more complicated procedure, where the scrotum is left intact, with the testes removed. This procedure must only be done by a vet.

In Britain, it is not considered acceptable to put a doe younger than eighteen months into kid, as she will not be physically mature enough. It would prevent strong bone and organ development and may shorten her life.

Goat pregnancies usually last around 150 days, but can be as long as 156 days or as short as 144. The nanny usually bears one to four kids, with an average of two. Her shape will change as the foetus grows and about ten days before the kids are born, her udder will enlarge. She must be regularly monitored and supervised by a veterinarian or an experienced breeder. It would be unthinkable for the amateur to act as the midwife, as there could be many complications that would result in the loss of both the kid(s) and the nanny.

Goat breeding will always raise the question of what you do with the kids. Should unwanted males be euthanased at birth, or castrated and moved on to loving homes (if you can find them)? Is there room on the property to keep more animals, or do they all need to move on to new homes? Do you really need to kid at all if you not planning to keep them? You should bear in mind that it is not always easy to find suitable homes.

If you want to keep a goat for milk (a Milch goat), it would better to obtain a nanny that is in milk, preferably a young animal, from a breeder. You should also ask for tuition in the actual milking technique. This can be quite daunting, and if done incorrectly may hurt the animal and damage the teats. Good care and attention must be given to the udder, which needs to be cleaned before and after the milking process. A cream should be applied to keep the skin soft, and prevent soreness.

MILKING

The kid(s) must have access to the colostrum milk (the first milk containing the antibodies) within half an hour of birth and should remain on this for 48 hours.

Hand rearing kids is time consuming and needs to be integrated into your other household chores.

No milk should be taken for human consumption at this time. The kid(s) can be removed from the mother after about two weeks and hand reared if the milk is to be used by the goat keeper. If large quantities are not required for human consumption, the kid could be separated from the nanny overnight, who would then be milked first thing in the morning, while the kid(s) are bottle fed. They would then be put back with their mother for the whole day. But both of these systems are very traumatic for the mother and her offspring, and will be very time consuming. Full time hand rearing with a bottle means that the babies need to be fed at least three times during the day, and once more late at night. Standard breed kids need to drink about four pints of milk a day, so hand rearing is difficult to integrate into the routine of a busy household.

When the kids are left with their mother, natural weaning takes place at about four to five months of age. But if you are rearing the kids with the bottle, you should gradually reduce the number of feeds after about 10 to 12 weeks so that the kids are ready for weaning at about 16 to 18 weeks. The teats should be treated with an anti-bacterial drip after milking to help prevent mastitus. An udder cream can be applied if there are any signs of soreness to the skin.

Goat Breeds

ZOOLOGICALLY, THE GOAT IS PLACED in the cloven-hoofed ruminant family of the Bovidae, "the horned ones", where one would also find sheep, cattle, and gazelles. Goats are part of a subfamily that includes sheep and other relatives such as the musk ox and the chamois, in which four species of wild goat make up the genus Capra.

Capra hircus: The Bezoar goat is found in the mountains of Western Asia and ranges from the Himalayas to the Greek islands.

Capra falconieri: The Markhor is found in Afghanistan and Pakistan, and is particularly attractive because of its twisted curled horns as opposed to the sabre-shaped horns of the other wild goat species.

Capra ibex: Although the Ibex and Nubian Ibex are from the mountains around the Red Sea and the Caucasian Tur comes from the Caucasus, the animals are sometimes considered to be the same breed.

Capra pyrenaica: The Spanish Ibex.

All of these wild goats are still in existence, but the Nubian Ibex is under the threat of extinction and the mountain goat survives only due to strict conservation regulations. Wild goat numbers in Greece are gradually falling and in the national park of Crete, their survival is entirely due to the protection of conservation laws. By contrast, uncontrolled hunting threatens to exterminate the wild goat in many regions of Asia.

Wild goats can be found in all mountainous areas, and their body-build and lifestyle are very well adapted to this environment. Many characteristics of this adaptation to mountain life can be seen in our domestic goats. These animals are certain to be descendants of the Bezoar goat Capra hircus. In central Asia, some domestic goats appear to be descended from the Markhor goat of Afghanistan and Pakistan.

A traditonal goat herd in Ithaki, Greece.

THE ENGLISH (OLD ENGLISH)

Prior to the twentieth-century importation of goats into the British Isles, the English goat was the most common breed. Although cross-breeding has diluted its bloodlines, many characteristics of this breed have survived in some animals. This means that, with careful selection, it is possible to breed back the genetic traits of this breed. The colours of the English goat vary from shades of brown or grey, with a dark line along their backs, down their necks, front legs, and flanks. Their coat is dense and short but has long fringes on the back and flanks. Thick tufts on the legs are also allowed, but neck tassels are not present.

The true Old English goat has quite a heavily built body, with a short tapering head. Its horns are carried wide apart, rising a little from the skull then pointing outwards in a curve. The coat colour is generally black or brown with white markings down the nose and on the legs, and there is shaggy hair around the back legs.

The Welsh goat.

IRISH

A little like the Old English, the Irish is a shaggy goat sporting a long beard on its large head, and has long, pointed horns which rise straight upward. The coat is often speckled with white hairs and can be black, brown, or grey.

WELSH

The Welsh goat is very similar to the Irish, but is a lot more graceful in appearance with a smaller lighter build. The black-necked Welsh goat looks very similar to the Bagot but without the shaggy beard. It has long hair that is black at the front end and white in mid and rear end. It is kept mainly as a meat animal.

In the early twentieth century there were just two recognised distinct breeds of domestic goat in the United Kingdom, the Anglo Nubian and the Toggenburg, which accounted for twenty per cent of the registrations in the British Goat Society's Herd Book. At present the BSG recognises eight different breeds of dairy goat, plus the crossbreeds.

ANY OTHER VARIETY

These are the crossbreeds that are grading up to a specific herd book; they often carry incorrect markings. However it is now possible, with careful selection, for cross bred goats to be upgraded, in time, to either Anglo-Nubian, British Alpine, British Saanen, British Toggenburg, or British Guernsey. Pure Saanen, Pure Toggenburg and Golden Guernsey are the exceptions to this rule, as they all have their own closed herd book that disallows crossbred upgrading.

The British Alpine, British Toggenburg, and the Toggenburg have white markings (referred to as "Swiss markings") down the nose, around the ears and tail, and on the legs. A crossbred goat or one that is grading up may also have these markings.

ANGLO NUBIAN

The largest of the British dairy breeds, the Anglo Nubian has its origins in the Middle East. The Zariby (or Nubian) goat, originally from Egypt, Abyssinia, and Jumna Pan, was brought to London at the end of the nineteenth and early twentieth centuries on P & O Line steamers. The animals were taken on board to provide the passengers with fresh milk on the journey to England. When the ships docked in London, the goats became very popular with English goat keepers. They bred them with indigenous British goats, resulting in a cross breed that became known as the Anglo-Nubian. The Anglo-Nubian is probably the most distinctive of the British breeds with its big convex "Roman" nose, and long pendulous ears. Its legs are straight, and long in proportion to the body depth. The legs are without the "cow hocks" tendency that appears in other breeds. The withers and hipbones are high, which gives a slight curve to the back. The head is held high, resulting in an upright stance, and a rather arrogant demeanour. Neck tassels are not allowed for registration purposes, but do often appear. The coat is very short and smooth, and can be of many different and attractive colours. These vary from plain brown with black and white markings around the head, to a multi or tri mixture of black, brown, and white markings with a patched or rosette formation. Unfortunately this breed frequently carries long swinging udders with large teats, which can prove difficult for the offspring to feed from. Compared with other breeds there is also a higher level of butterfat and protein in the milk, which helps in the development of cheese yield and is excellent for yoghurt and butter making. Latterly, the Anglo-Nubian has also become very popular in Canada and the USA. But despite all its attributes, I would not recommend this breed for the beginner, as they can be difficult to handle particularly when they are being stubborn.

The Anglo-Nubian goat.

SAANEN

A few Saanen goats were imported into Britain from the Saane and Simmental valleys of Switzerland at the turn of the twentieth century, but the majority came into the country in 1922 and then in 1965. In appearance, the pure Saanen is white with short legs in proportion to depth of body, unlike the British Saanen, which has longer legs. The back should be straight without a hump (this is referred to as a "roach" back) with a short, smooth coat; sometimes there is also a fringe along the spine. The face is slightly dished with erect ears pointing forward. Saanen milk yields are generally good. The breed is probably the most popular dairy goat throughout the world, in both its pure and crossbred form.

BRITISH SAANEN

This is the most popular breed in Britain. Its ancestors were imported Saanens which were crossbred with indigenous British goats in the early twentieth century. This crossbreeding produced a larger dairy goat with a high milk yield. In appearance, the British Saanen looks like its pure ancestor, but has longer legs and a straighter facial line. Its coat is white, and sometimes has biscuit coloured patches, which moult out in the spring. Like the Saanen, the British Saanen sometimes carries a spinal fringe. A

A pure Saanen milk goat.

good-natured breed, it is usually the one chosen for commercial milk production.

TOGGENBURG

The importation of another Swiss goat breed, the Toggenburg (from the regions of Obertoggenburg and Werdenburg) began in the late nineteenth century. But the majority of Toggenburgs arrived with the Saanen imports of 1922 and 1965. Very tight importation regulations now prevent further imports. The 1965 intake did nothing to improve milk yields, and other breeds took over in popularity. But Toggenburgs do well in other countries, so the problem is thought to be in the gene pool of the British stock rather than with the breed itself.

In appearance, the Toggenburg is a small, sturdy, compact animal that thrives on extensive browsing and grazing. The colour varies from a silvery fawn to brown with white "Swiss markings" and there is nearly always some long hair on what might otherwise be a smooth coat; this long hair may be just a fringe down the back and hind legs, or might cover the whole body except the neck and head. The face is slightly dished and the throat may or may not carry tassels. This good-natured, affectionate little animal is about four inches shorter than the Saanen and handles well. It is also an economical milker and may suit the beginner who wants a medium-sized breed. Toggenburgs also make excellent pets and are very popular in both the USA and England.

The Toggenburg goat.

The British Toggenburg

BRITISH TOGGENBURG

In the early twentieth century, pure Toggenburg males were mated with animals of mixed descent, and by 1920, their descendants were recognised as a distinct breed. They were given a place in the breed section of the BGS herd book in 1925. The British Toggenburg is larger than the pure Toggenburg, and varies in colour from fawn to chocolate brown with "Swiss markings". Its coat is short and fine without fringes of long hair. Unlike its pure ancestor, the British Toggenburg has a straight face and usually produces high milk yields. It is a very popular breed for the milking parlour.

The Alpine goat.

ALPINE

Alpine goats are named for their origins in the mountainous regions of France and Switzerland, and are very like the Saanen in shape and milk production but are more robust. The coat colour is usually mostly fawn as they are descendants of the Swiss chamois-coloured goat, but the breed also has black and variegated individuals. This breed is not found in Britain.

BRITISH ALPINE

This breed is a British creation and there are no Swiss or French equivalents. In 1903, a French Alpine goat named Sedgemore Faith was imported into England and she and her progeny are believed to be the antecedents of the present day British Alpines. In 1919, after sixteen years of carefully selective crossbreeding, the breed was defined as the British Alpine and a breed section was opened with the BGS.

Sedgemore Faith had a striking black coat with white "Swiss markings " and a good contemporary British Alpine will have this colour presentation in its short coat. The face line can be dished (concave) or straight, and the neck can present with or without tassels The legs are long, making it quite a large rangy animal to handle, but the breed was very popular for many years because of its high milk yield. The breed's popularity declined in the 1970s, as it became more and more difficult to produce animals with good udder formation and coat colour. This was because there were no pure breed bloodlines to re-enhance and maintain the breed characteristics. This has resulted in British Alpines having a reputation for being difficult milkers, due to their unmanageable udder formation.

GOLDEN GUERNSEY

This breed was developed in the Middle Ages. It originates from Syrian goats that were traded off Mediterranean sailing ships, and written about by Herodotus (c.485-425B). I quote: the animals have "wondrous ears turn upwards and outwards at the tips in tribute to Apollo who gave them their golden coats". The first reference to the breed being found on the island of Guernsey was made in a guidebook of 1826.

When Miss Miriam Milbourne of L'Ancresse moved to Guernsey in 1924, she found that the breed was almost as extinct as the Golden Donkey. For the rest of her life, she endeavoured to save the

The Golden Guernsey is named for its distinctive golden coat.

Golden Guernsey by taking them out of the scrub herd in which they roamed to selectively breed them, and record their bloodlines. During the wartime German occupation, Miriam gathered up some bucks and as many does as she could find and hid them in caves. This was to prevent them being hunted and eaten when food supplies became short. The breed has survived from these few rescued animals. But due to its very low numbers and limited bloodlines, this lovely goat is now among many threatened species of farm animals under the protection of the Rare Breeds Survival Trust.

The most outstanding feature of the Golden Guernsey is its true golden colour that can vary from a shade of pale tea through to a ginger, in both coat and skin. Both colours are acceptable with or without a few white markings but no "Swiss markings" are acceptable in the breed. The hair can be either long or short. The Golden Guernsey has a small stature, with strong, fine bones. Its neck is long and without tassels, its face can either be slightly dished or straight and, as Herodotus had written, its ears turn outwards and upwards at the tips. It is hardier than the Swiss dairy breeds and produces an adequate quantity of milk.

This delightful loving and loveable animal is becoming more popular as it makes a really good pet.

ENGLISH or BRITISH GUERNSEY

This breed was created out of a cross of Saanen males and Golden Guernsey females that had been brought to the mainland in 1965. The resulting animals were then crossed back with Golden Guernsey and English Guernsey males for three generations. It has the same characteristics as the Golden Guernsey but is often slightly larger and paler in colour. Unlike its pure ancestor, neck tassels are permitted. The BGS opened a register for the breed in 1975, and breeders endeavour to improve the productivity and conformation of the breed with the use of British Saanen bloodlines.

THE ANGORA

The Angora is one of the oldest breeds of goat in the world, having been mentioned in the bible (Exodus: Chapter 35, Verse 26). It comes from the highlands of Turkish Anatolia, and its name is derived from the city of Ankara. The breed is kept for its coat of luxurious mohair, which has a beautiful lustre and is very hard wearing. To the textile industry, mohair is known it as "the diamond fibre". To maintain top quality mohair, Angora goats need a good diet high in protein. The breed is generally smaller and finer boned than the dairy breeds. It generally has a white coat but can sometimes have a dark blue/grey colour. The coat is made up of waves and ringlets of hair that grows continuously at a rate of about an inch a month. The breed is horned and Angoras often have turquoise blue eyes. The very best hair comes from young goatlings, but it must be removed from goats of all ages at least twice a year. This is because of the coat's rapid growth, and the risk of lice and mite infestation on the skin. Shearing usually takes place in early spring and mid-autumn unless the animals are to be exhibited, in which case shearing will be much earlier to ensure a good display of hair at the shows. There was a time in the late twentieth century when keeping Angoras was fashionable as a money making hobby, and people spent thousands of pounds purchasing very expensive animals. Unfortunately, the price of mohair fell dramatically, and the animals were moved on to become pets or meat.

Angoras are still found in their native Turkey, but are primarily located in South Africa and Texas, USA. But there are also many examples of the breed in Britain, France, and Germany. Strict veterinary regulations now make it difficult to procure

Two Angoras show off their curly coats.

animals from the traditional breeding areas.

The Angora is a very likeable and agreeable animal, but is a lot of work for the novice to undertake.

THE BOER

The Boer originated in South Africa and is primarily a meat-producing breed of goat. Some castrated Boer males weigh as much as 220 pounds (100kg). It could be mistaken for an Anglo Nubian but the body is much deeper and wider, with shorter legs and neck. The Boer's nose is also straighter and shorter than that of the Anglo Nubian, while their ears are also floppy but smaller. Boers come in many different colours but breeders prefer the combination of a white body with a chestnut brown head. Another look alike breed is the Shami or Damascus goat. In 1988, the first Boer studs were established in Britain. They came by way of France, Germany, Ireland, and Israel. Due to its size and strength, the Boer is not a goat for the beginner.

THE BAGOT

To look at, the Bagot is very similar to the Swiss Schwarzhal goat, with its black head and shoulders and white body. It is thought that, in the twelfth century, King Richard I was given a herd of beautifully striking goats on his return journey from the Crusades. In the 1390's, King Richard II presented the royal herd of goats to Sir John Bagot of Staffordshire as a token of his appreciations for a good days hunting at his host's estate. The goat was subsequently incorporated into the Bagot family crest, and the herd continued to live in a semi-wild state in Bagot Park until the late 1970s. At this time, the park was sold and flooded for a reservoir. The survivors of the Bagot herd were rounded up and presented to the Rare Breeds Survival Trust by Lady Nancy Bagot. The Trust then re-homed the animals in many suitable homes around the country to ensure the survival of the breed.

Bagots are light boned and small to

The Boer goat.

The Bagot goat.

medium in size. They have beards and long curving horns. These can sometimes meet at the back and will need trimming. The animal's coat is long and white from the shoulder area with a black head and forequarters (the throat and shoulders). Many animals also have spots and patches of black on their hindquarters and small white facial blazes. At present, these features are accepted but regarded as faults. In time, these faults will be bred out as breed numbers increase and genetic improvement is engineered. Bagots are a non-productive breed, but are very ornamental, delicate, timid and very attractive. They are still on the Rare Breed Survival Trust register as a threatened breed.

THE PYGMY or WEST AFRICAN DWARF

As the secondary name suggests this delightful little goat originates from Africa where, in the forest areas of West and Central districts, it is an integral part of the inhabitants' existence. It is a prolific breeder, and is kept as a meat animal. The breed's milk production is only enough to feed the offspring. It is a small compact animal with a short head and neck, and sturdy legs that look unrelated to the body length. Any colour or marking is acceptable but a purist breeder would reject a pure white coat and "Swiss markings" on the face. The breed is genetically horned. Although disbudding is permitted, the author does not approve of the act of removal. This delightful little breed is very popular throughout Europe and North America as it makes an ideal family pet and a good companion to other species.

A Pygmy goat

OTHER BREEDS IN THE WORLD

CASHMERE

Cashmere goats have been bred in Central Asia for their fine hair for hundreds of years. They first came to the notice of Europeans during Napoleon's time. In the twentieth century, Australian scientists instigated a breeding programme of feral goats, aiming to produce high levels of cashmere hair. By the 1980's, this programme proved so successful that animals were being exported to various parts of the world. Scientists in the USA began a similar experiment using the Spanish feral goat, and like their Australian colleagues, they were soon exporting animals to places like Canada. The Canadians used goats from both programmes to develop their own cashmere-producing animals. The hair colour of this breed can be pure white, cream, soft grey, or brown. The goat is usually lightly built, not unlike an Angora, but the Spanish feral type is heavier. These animals are bred as meat producers in addition to being kept for their hair. Given their wild and feral origins, these goats are all genetically programmed to grow horns.

The fine undercoat or down which can be present in most dairy breeds is also called cashmere hair. But their short coat produces only a small amount of hair. By contrast, feral goats that live in the wild grow much more under hair than the cosseted, domesticated animal. Many of the feral goats that live in various parts of the British Isles can be regarded as cashmere-producing goats. The significant amount of cashmere hair that they produce is as good as any in the world.

KIKO

This large framed, hardy goat comes from the mountainous areas of New Zealand. The Kiko has a reputation for making good weight gains without supplementary feeding. It is usually white but can carry colour genes. In winter, its coat is long and flowing, but it is smooth in the summer. The bucks are bold and horned the females are very feminine, with good udder formation. The Kiko is an ideal breed for meat production.

The Kiko is a large-framed, hardy goat.

SHAMI or DAMASCUS GOAT

This is a large breed from the Near East. It looks a little like the Anglo Nubian goat, and may well be one of its ancestors. It has smaller hanging ears and longish hair at the rear. The breed comes in various colours. Its milk production is good.

GENETIC FAULTS

For breeding purposes the following genetic faults, which can be found in all breeds, are not acceptable:

Mouth or jaw, over shot, under shot, or twisted

If this defect is very obvious at birth the kid may well be euthanased. But if the defect is not too pronounced, so that the animal could eat normally, then the goat could live a normal, happy life. We have had both sheep and goats of this disposition.

Testicals, not properly descended or fewer than two present

We have taken in many "wethers" only to find that they were not, but that their testicles were retained. These require surgical removal by the vet.

Teat defects of any kind: fish tail (divided) teats, more than two teats present

If you do not want to breed from a goat, then a deformed udder would not be a disaster. But if it drags on the ground, surgery would be required. Rejected breeding animals of this kind may well be the pet that you would love to look after, so talk to a breeder to see what is available.

LEGAL REQUIREMENTS FOR GOAT KEEPERS

The Animal Welfare Act 2006 makes the owners and keepers of goats responsible for the welfare needs of their animals.

The Welfare of Farmed Animals (England) Regulations 2007 covers the welfare of commercially farmed goats and there is similar legislation in Scotland, Wales, and Northern Ireland.

DEFRA also issues Codes of Recommendations for the welfare of Livestock. Goat keepers, or those responsible for goats, should acquaint themselves with these codes. They are available from DEFRA.

THE IDENTIFICATION AND MOVEMENT OF GOATS

In an effort to prevent the spread of disease in the UK there are strict laws governing the identification of goats, the premises on which they live and the movement of animals between premises. These rules apply equally to a single pet goat or a commercial herd of goats.

1. A holding number for the land on which the goats are to be kept must be obtained.
2. The goats must be registered with DEFRA.
3. You should keep a register of your animals on paper, or on the computer.
4. You should ensure that your animals are individually identified, according to the current legislation.
5. You will need to obtain the appropriate licence for any movement of your goats between properties, according to the current legislation.

In the UK, all of these requirements must be fulfilled before the animals are brought onto your property, for the first time and for any subsequent moves.

DECEASED ANIMALS

Goats are classified as farm animals and must not be buried or cremated on the goat keeper's premises. Carcasses should be disposed of through an approved handler. The Animal Health Department of your local Trading Standards Office will advise you of authorised outlets if your friendly goat keeper cannot do so. Their telephone number should be obtained and kept safe with the number of your vet.

MILK AND MEAT PRODUCTION

Please be aware that if you are rearing goats for meat, there are strict laws relating to animal welfare when they are slaughtered. There are also rules governing meat hygiene, and meat inspection. Further details are available from DEFRA, the Humane Slaughter Association, and the Food Standards Agency.

If you are rearing goats to supply milk to others, please be aware that there is legislation relating to milk hygiene. Details available from the Food Standards Agency.

NOTIFIABLE DISEASES

If goat keepers suspect Scrapie or Foot and Mouth Disease then a veterinarian must be contacted and DEFRA advised.

MEDICINE RECORDS

A record of all medical products prescribed by a vet must be kept in accordance with current legislation.

USEFUL CONTACTS FOR GOAT KEEPERS

Department for the Environment, Food, and Rural Affairs (DEFRA)
www.defra.gov.uk
Telephone 08459 33 55 77

BRITISH GOAT SOCIETY
www.allgoats.com
Telephone 01626 833168
B.G.S. Secretary, 34-36 Fore Street, Bovey Tracey, Near Newton Abbot, Devon TQ13 9AD
County branches of the BGS are listed in the telephone directory.

GOAT WELFARE
This is a useful source of information and advice.
www.bushpark.co.uk/goatwelfare/goatwelfare.htm
Telephone 01409 221230 or 01822 820412

GOAT VETERINARY SOCIETY
www.goatvetsoc.co.uk
Telephone: 01531 820074
G.V.S. Secretary
29 Winfield, Newent, Gloucestershire GL18 1QB

R.S.P.C.A.
www.rspca.org.uk
Telephone 0300 1234 555
Fax. 0303 1234 555
RSPCA,
Wilberforce Way, Southwater, Horsham, West Sussex RH13 9RS

RSPCA Farm Animal Department
www.rspca.org.uk/farmanimals
E-mail: farm_animals@rspca.org.uk

Acknowledgements **Buttercups Goat Sanctuary**, www.buttercups.org.uk. **Debbie @ Shanvern Goats,** British Toggenburg, www.elliesdairy.co.uk. **Forsham Cottage Arks**, www.forshamcottagearks.co.uk. **Jan Newell,** Newell's Kiko Farm, kikosrus.com. **Sara Dzimianski**, Violet Vale Saanen Diary Goats. **Alpine Goat** Connie, Little Dixie Farm. **Bagot Goat** Dave Brown, Wainright Walks. **Pygmy Goat** Rise Wilson. **Boer Goat** Adrianne Bell. **Welsh Goat** D.Sallery.